Global Pets Community BV
PO Box 1719
3800 BS, AMERSFOORT
NETHERLANDS

phone: +31-33-4225833
email: info@globalpets.community

Content

KVK: 54632730
BTW NL8538.04.849.B01

info@globalpets.community

Global Pets Community BV
PO Box 1719
3800 BS, AMERSFOORT, NL

Introduction

All work serves a purpose. Hence also the output of R&D, in a commercial environment, should serve the purpose to contribute to the business results. This is not nebulous but rather concrete.

The difference with the output of a product manager is that R&D is usually not involved in direct sales. However, R&D constitutes the foundation, the pillars, on which Marketing & Sales can establish their communication strategies.

Whilst the role of Marketing & Sales is to construct a proposition, which is surprising, meaningful, desirable and delightful to consumers, they cannot do this without the deep involvement of R&D. Indeed, R&D owns the scientific, nutritional and technical knowledge, which constitutes the base of product uniqueness and nutritional adequacy. Not the least R&D also provides the substantiation for credible claim development.

In your role as a product manager it will be key to create an enriching, smoothly running working relationship with R&D. This course will help you in setting your first steps on how to realize this.

KVK: 54632730
BTW NL8538.04.849.B01

info@globalpets.community

Global Pets Community BV
PO Box 1719
3800 BS, AMERSFOORT, NL

Page 4 of 23

1. Understanding the role and position of R&D

Why is R&D in your organization is organized the way it is? What is the role and primary function of R&D, the desired output? And what is the importance to work within a framework of mission, vision, and overall strategy? In what way are the company's beliefs and, for instance the nutritional credo an overarching guidance? Should you look at R&D as a department residing in an ivory tower, rather detached from the rest of the organization, at times even unapproachable… then this chapter will hopefully change your mind.

Indeed: the overarching objective of R&D in a FMCG (Fast Moving Consumer Goods) environment (like pet food) is to use their expertise to make business thrive. This does not necessarily mean that R&D people are involved with direct sales; there are many other aspects of work-output that ultimately will lead to long-term business results and on which R&D has a unique influence. Let's discover together what those aspects are.

1.1 Place in the organization

The company's business model and strategy defines the 'nature' of the company you joined. Depending on the company's 'nature', R&D can have many forms and can be organized in several different ways:

- It can be 'non existent' if you joined a company that is purely distributing brands of other (producing) companies. In this case R&D resides with the producing company who is also the brand-owner. In your role, as a product manager, you will probably seldom be in direct contact with R&D, rather you will convey to your customers and target consumers the message that the brand owner determined to be the most suitable one to sell the brand in question. How that message was created is not necessarily your concern, as long as you receive the ammunition (arguments) to do a good job.

- Quite opposite to the above, R&D can be 'in-house' as a full department, with many employees and many different functions within R&D. In this form, R&D may claim a sizable chunk of the yearly budget of the company. On occasion it can also be just a person doing desk-research. Or the R&D function can be outsourced in all or in parts of its deliverables. Very often R&D has links to universities, research centers, external think tanks, etc. In these cases we sometimes talk of 'collaborative research' between the company and the external instances.

KVK: 54632730
BTW NL8538.04.849.B01

info@globalpets.community

Global Pets Community BV
PO Box 1719
3800 BS, AMERSFOORT, NL

Page 5 of 23

1.2 Impact of R&D

The nature of the company you joined may also dictate the impact of R&D.

It is important for you to understand that a company can be R&D steered, hence playing a leading role in the product offer, but it can also 'serve' the purposes of Marketing & Sales being in the driver's seat. Between both extremes many layers of grey exist though…

When R&D is steering the primary focus could be on the nutritional needs of dogs and cats. When Marketing is steering, consumer trends and consumer wants may prevail. Both extremes might not lead to long lasting sales results, and, as usual, wisdom is to create balance between both models. A well-defined 'marketing mix' can sort things out on this front at least when 'walk the talk' is high on the radar screen of the company and all it's employees.

1.3 Mission and vision

Just like for any other function in a company, being 'on strategy' is crucial for R&D's output. What does that mean? Each company should have a *mission statement* (what is their reason to be in that particular business) and a *vision statement* (a description of a plausible future, desirable for all).

KVK: 54632730
BTW NL8538.04.849.B01

info@globalpets.community

Global Pets Community BV
PO Box 1719
3800 BS, AMERSFOORT, NL

Page 6 of 23

It is of key-importance for these two statements to be the guiding principles of what is done… and what will not be done.

Example: if the mission statement is: "to enhance the well-being of dogs and cats", then no products should be developed that contain elements that do not contribute to deliver just that! The whole food but also each ingredient and additive should be defined in function of whether they do or don't contribute to the realization of the mission: to enhance dogs' and cats' wellbeing.

Same with the vision statement: if this would be "to become the world-leader in dog and cat nutrition" then innovation is the overriding goal to pursue. New concepts in nutrition, contributing to the wellbeing, should be discovered, implemented and shared with the world. Why 'shared'? Because you might *be* the world-leader, but if the world does not recognize you as the leader it does you no good business-wise.

1.4 Nutritional credo

In order to achieve the above, you as a Product Manager, will have to thoroughly understand the nutritional credo of the company. This is the set of nutritional principles, the beliefs on how dogs and cats should be best fed, that the company will systematically apply, in order to develop products and services for the fulfilment of the mission, the vision… and the business results. The nutritional credo will keep all employees on a steady course. It will be a beacon and avoid that distraction will occur by giving in on at times silly market trends or copying competitive aberrations about decent dog and cat nutrition.

With this comes also the claim substantiation. For long-term credibility sake you cannot afford yourself to claim features and benefits about your products, which cannot be substantiated by sound evidence and/or scientific back up.

R&D can and should help you with this. Your R&D colleagues will help you to avoid pitfalls related to overly enthusiastic marketing messages. They will help you to become more confident in what you can claim and, at times, show you ways to communicate that you had not discovered so far. Finding communication opportunities so to speak.

Does it become obvious to you that, given the above, R&D has also a major role to play in external communication? That you as a product manager should partner with your colleagues from R&D to receive the ammunition for establishing your communication tools content wise? That R&D is your resource to better understand the uniqueness, features and benefits of your products and those of your competitors? You will also need R&D to establish a concept that will be of paramount importance in your job as a product manager: the creation of a sustainable competitive differentiation.

1.5 Functions of R&D

Depending on the structure and size of your company, R&D can and should also play an active role in building credibility (company and product equity) for your brands and products by participating in seminars, congresses, scientific articles, etc. This will especially be the case if your R&D department is involved with fundamental research.

KVK: 54632730
BTW NL8538.04.849.B01

info@globalpets.community

Global Pets Community BV
PO Box 1719
3800 BS, AMERSFOORT, NL

Page 7 of 23

Major functions within the R&D department that you will most likely be involved with are:
- Product development/formulation
- Regulatory
- Quality control
- Technical communication
- Testing facilities
- The pilot plant

In other courses for the product manager, we will go into further detail concerning these functions. Also shall we highlight what basic skill-set you should acquire in order to work effectively with R&D. This skill-set includes:
- Basic nutrition
- Product knowledge
- Competitive differentiation
- Regulatory framework on marketing and claim development
- Work flow between Product Manager and R&D

KVK: 54632730
BTW NL8538.04.849.B01

info@globalpets.community

Global Pets Community BV
PO Box 1719
3800 BS, AMERSFOORT, NL

Page 8 of 23

2. Major functions within an R&D department

From the previous chapter you have understood that R&D can have many different forms and output, depending on the company it is embedded in. What we will discover in this chapter is how a 'full swing' R&D department can be organized, or – in a more practical way – we will discuss what the major functions are and how their output can contribute to overall business. More importantly for you personally, you will learn what the benefits are for you to have a thorough understanding of the potential that R&D can offer you in your daily work.

Again, we consider in this chapter that R&D is fully developed in the company. Of course you will have to adapt the information in this chapter according to the situation in your company.

2.1 Senior management
To start at the top: the R&D manager. This person has to provide direction to all what the department is doing (and not doing) and has overall responsibility of the final outcome of all activities of all employees, and of the efficiency and effectiveness of the R&D as a whole.

The senior manager is also responsible that the money (budget) allocated to R&D benefits the overall business in a maximum way. In order to do this it is obvious and of paramount importance that R&D's senior manager is part of the management team of the company and attends, discuss, influence and (above all) understand fully the company objectives and the major strategies to achieve the objectives. Only in this way can the Senior R&D Manager define what the R&D department's priorities should be in order to apply the company strategy and define the R&D strategies, plans and programmes to achieve this. In normal circumstances you will have little or no direct working relationship with this senior manager.

2.2 Fundamental research
This person or team within R&D might come across as very 'theoretical'. This team is indeed very much involved with all kind of studies to enhance the knowledge of the physiology of pets, and how nutrition can impact the maintenance or improvement of the physiologic processes. Very often they are involved with pathology (diseases) of animals and they try to find nutritional tools as part of the prevention or treatment of these diseases.

The discovery of the role of 'moderately fermentable fibre' is a good example of this. In the recent past common understanding was that fibre was indigestible bulk whereas part of the fibre can be fermented in the large intestine by the bacteria. The fermentation products (short chain fatty acids) improve the effectiveness of the large intestine (see the nutrition course).

Another example is the role of omega 3 fatty acids in modulating inflammatory responses in the body and the importance to have an optimal balance between omega 6 and omega 3 fatty acids in the diet.

Both examples have been the result of fundamental research that often is also done in cooperation with universities and research centres (it is then called 'collaborative research').

KVK: 54632730
BTW NL8538.04.849.B01

info@globalpets.community

Global Pets Community BV
PO Box 1719
3800 BS, AMERSFOORT, NL

Page 9 of 23

But their work is not limited to physiology of pets. Fundamental research can also be practiced on many other topics: new technologic applications or new ingredients, new additives, to name just a few.

2.3 Product development and formulation

This team implements the outcome of fundamental research into new product ideas and/or improvement of existing products. They work with a vast knowledge base on ingredient specifications. They know all the ins and outs of the nutritional value and pitfalls of ingredients. They also know about the true nutritional needs of the pets they are focusing on and are therefore able to define a 'nutritional matrix' based on those needs. They can play with several scenarios for the formulation of new products or improvement of new products.

Of course they do this in close cooperation with marketing and sales, as the product needs to be able to be sold. Therefore the criteria for product features and benefits, the price, the shape, the colour, and many other aspects come in play. Most important is whether the new product will have a sustainable competitive differentiation and whether the key benefits are meaningful to consumers (and their pets).

Needless to say that because of their close cooperation with marketing and sales on new concepts and product upgrades, this team will be one of your close partners in your job as a product manager.

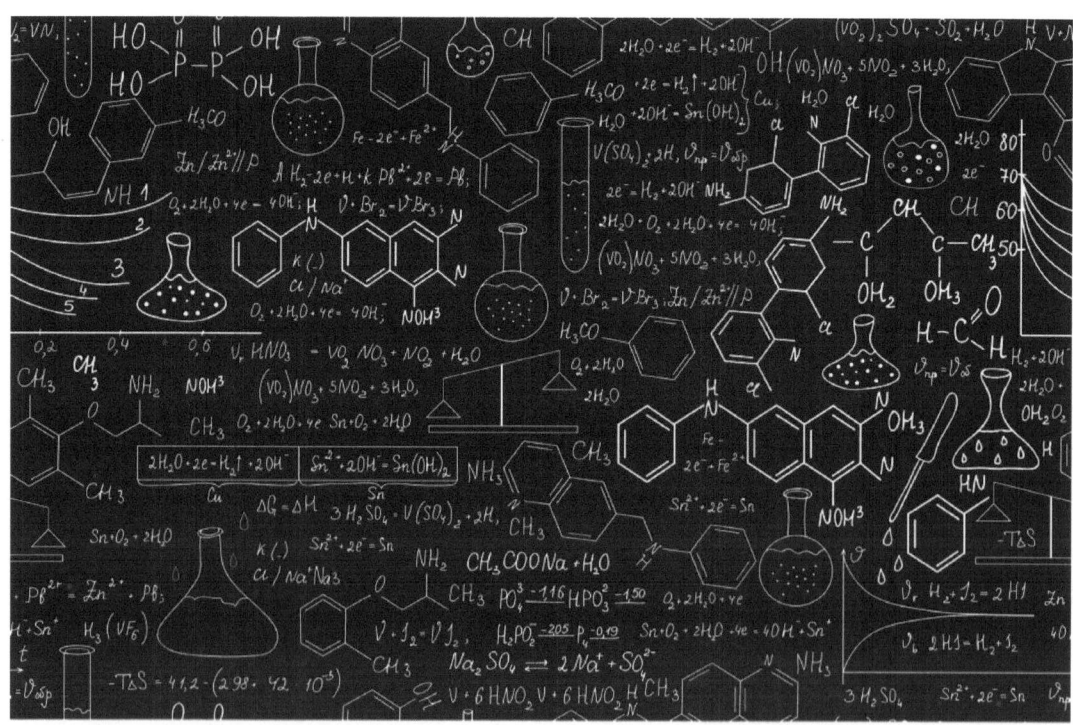

KVK: 54632730
BTW NL8538.04.849.B01

info@globalpets.community

Global Pets Community BV
PO Box 1719
3800 BS, AMERSFOORT, NL

Page 10 of 23

2.4 Regulatory department

The production and marketing of compound pet food is strictly regulated. The role of this department is to make sure that all what is said about a product is CLADD. This stands for Consistent, Legal, Accurate, Demonstrable and Defendable.

Consistent: it is key that all internal and external parties involved with the product in the market tell the same story. Would it not be disastrous for the credibility of the company and the brand if a company sales person, a distributor or retailer would 'create' on it's own product benefits that are false, untrue or vastly exaggerated? Or starts to make medicinal claims? The role of R&D is to be the watchdog that such a thing won't happen.

Legal: all what is said about the product, the label, the website, the brochures, etc., falls within the framework of the legislation. Hence their role is predominantly preventive. But at times they will have to go and defend the company's position for official instances or competitors. Does it not make sense that Regulatory should have an eye on the ball with your work as well? Yes it does and you will have many encounters with Regulatory.

So in a nutshell the output of Regulatory will be to safeguard that the product and the communication stays within the legal framework, establish the claim substantiation files, defend the company position in case of litigations. But they also play an important role in monitoring competitive claims and products and make sure that competitors don't take advantage in a non-legal way.

At last he had found the Regulatory Guidelines.

2.5 Technical communication

This is usually a team of persons with 'technical' background (veterinarians, biologists, engineers) with strong communication skills. They convert the 'facts and figures' from above teams and work in close relationship with marketing and sales on a communication plot.

They are most suitable to translate the features and benefits of a product into a credible, sustainable message understandable and resonating with the different target groups in the market place. This is usually also the team that provides technical, nutrition, product training to the organization. They also will help you to understand the strengths and weaknesses of competitive products. This is also the team that will help you to answer consumer questions and at times to handle complaints when the food is suspected by consumers to create disease or death with an animal.

KVK: 54632730
BTW NL8538.04.849.B01

info@globalpets.community

Global Pets Community BV
PO Box 1719
3800 BS, AMERSFOORT, NL

Page 11 of 23

2.6 Testing facilities

Sometimes companies have their own kennels and catteries where they can test their (new) or existent products. And of course where competitive products can be tested and compared to your own products.

Feeding tests include tests for palatability or acceptance (first bite and total consumption), digestibility (stool quality). These feeding tests can also be performed in extern kennels or in-home environment. Pet food companies nowadays seldom practice so-called 'invasive' research. This means that no physical discomfort or euthanasia is performed in order to study the impact of nutrition on pets. However, normal diagnostic procedures, like taking blood or urine samples, are not considered being 'invasive' research. The same would happen if you go and see a vet with your pet.

> *Example:*
> *Taking blood samples as part of a feeding test can be important in order to determine the blood glucose level after a meal. Some ingredients, like rice, are digested very rapidly. The consequence is high blood glucose levels shortly after the food intake. High blood glucose levels create an 'insulin response' (the secretion of high amounts of insulin from the pancreas). Permanent feeding with high amounts of rice might on the long-term lead to the onset of diabetes.*

2.7 Pilot plant

Manufacturing facilities need to run efficiently in order to reduce cost to a maximum. One of the ways to achieve this, is to have 'long runs' (production of great volumes of the same product formula). Short runs and frequent changes of formula in the production line increase cost because time is lost for changes and cleaning between productions of two different recipes.

For this reason the manufacturing unit is often reserved for the production of the existing 'sellable' products, while a smaller-scale, more flexible pilot plant, can be dedicated to run small runs. This is very important when R&D is exploring various options in formulation in view of new product development or product upgrades/improvements.

2.8 Central laboratory and quality control

It is standard practice that all incoming material, before it goes into the production line, should be analyzed for its wholesomeness. The analysis may include checking the nutritious content, but also moisture, mould, pollutants, contaminants like insects or parasites, rancidity, foreign material, etc. Finished product also needs to be checked according to a strict scheme of sampling in order to make sure that the product produced is corresponding to the previously set criteria. Only then can the product be released from the factory.

By the way, the legislator demands that these samples be stored for a certain amount of time after production. This can also come in handy in case of product complaints from the field, as now the well-identified samples can be retrieved based on the batch codes on the package and either validate or deny the complaint.

KVK: 54632730
BTW NL8538.04.849.B01

info@globalpets.community

Global Pets Community BV
PO Box 1719
3800 BS, AMERSFOORT, NL

Page 12 of 23

Competitive products are also analyzed on a regular basis in view of monitoring your competitors.

A well-equipped laboratory is very expensive. Some machines cost a fortune. It is therefore important that these machines are used on a regular basis to justify a return on investment. For this reason part of the analysis can be outsourced to external labs while the basic analysis required for quality control (the Weender Analysis) is performed in-house.

In the next chapter we will provide some guidance on how the workflow between a product manager and R&D could look like. Later on, we will also see what basic skills you need in order to be successful and to take full advantage of your working relationship with R&D.

KVK: 54632730
BTW NL8538.04.849.B01

info@globalpets.community

Global Pets Community BV
PO Box 1719
3800 BS, AMERSFOORT, NL

Page 13 of 23

3. Positioning and defending the image of the company

Creating company-, brand-, and product equity is a marketing responsibility. But it is wise to integrate all functions and all individuals of the company in realizing this. It is indeed of crucial importance that the end-customer has a deeply entrenched positive image of not only the product, but also of the way the company and the employees behave. We will now discover how R&D can play a crucial role in positioning (and at times defend) the company's credibility.

We start from the perfect scenario where the company has a full-swing R&D department. Of course the potential of R&D in relation to equity creation will have to be adapted according to the reality of your company's organization.

3.1 Credibility at the base: innovative research

Perhaps the most specific task of R&D is to obtain new insights in how to feed dogs and cats in optimal ways. By doing so they would discover new applications on how nutrition may impact the physiology of pets. To realize this R&D can do in-house research. More often the innovation is the result of close collaboration with universities and/or other research institutions. We call this 'collaborative research'. This requires a highly competent staff at R&D and clearly defined projects with researchers at universities, who hereby also receive funds to perform their role. The findings from this research would then be applied into new or upgraded pet food products.

Collaborative research is expensive and therefore the projects need to be very carefully selected in order to increase the chance of having a commercially exploitable outcome.

The work of R&D is building company equity by the fact that:
- innovative products are brought to the market;
- new scientific facts are published, which results in a better understanding of the physiology and how nutrition can improve animal well-being;
- wrong concepts, and at times wrong vested opinions, are corrected;
- in sharing new discoveries with the external world the company is regarded as a scientific and nutritional leader.

Recognition of nutritional leadership is often realized with professionals (veterinarians, breeders, research-and teaching staff at universities). It is however important that the rest of the community also recognizes this leadership. Not in the least the end-customer. Bottom-line: you can *be* the leader, but if the rest of the world doesn't *recognize* you as a leader it won't help you much.

3.2 Competitive intelligence

Purpose: to have, at any time, a global and updated analysis of all aspects of our competitive environment.

Key assignments: to provide information on request on the following topics:

KVK: 54632730
BTW NL8538.04.849.B01

info@globalpets.community

Global Pets Community BV
PO Box 1719
3800 BS, AMERSFOORT, NL

Page 14 of 23

- Generalities
  - history
  - mission, vision, beliefs
  - philosophy

- Resources
  - people, organograms
  - main assets
  - financial and market figures: turnover, volumes trends, market shares

- Marketing
  - advertising campaigns
  - main claims and baselines
  - packagings
  - last promotions

- Products
  - products ranges (channel distribution, positioning)
  - products' technical characteristics (average analysis, ingredients lists, feeding instructions)
  - prices' lists

- Systems
  - plants: locations, capacities and main characteristics
  - suppliers
  - logistics

3.3 Spreading the word

Communication is at the base of equity building. It is therefore important that representatives of R&D make regular public appearances in order to share the knowledge. It is not a good strategy to keep everything close to the chest, rather R&D's sharing of their findings proves the intend of a company to take its share in improving the well-being of pets.

Often there are communication specialists at R&D who are not only highly scientifically qualified but who have also excellent communication skills. These technical communication managers will often work with you and marketing & sales in general to assure accuracy and consistency of the technical message. Their work will reflect in the quality of all kinds of communication tools: brochures, websites, in-store communication, etc.

The technical communication manager is often the person who translates the R&D findings into 'digestible' information, useful and meaningful to the various target audiences and exploitable in the market place.

KVK: 54632730
BTW NL8538.04.849.B01

info@globalpets.community

Global Pets Community BV
PO Box 1719
3800 BS, AMERSFOORT, NL

Page 15 of 23

It will be important, when you develop the packaging, the brochures and all other forms of collateral material, that there are feedback loops with the *Regulatory Affairs Manager*. It is obvious that his/her role is to make sure that all what is written and said about the product features and benefits is CLADD. This stands for:
- Consistent
- Legal
- Accurate
- Demonstrable
- Defendable

Both R&D functions will also help you with better understanding the strengths and weaknesses of competitive products and competitive vulnerabilities on both the technical and the regulatory side.

They will help you in answering customer or competitor's questions and complaints, especially when these questions and complaints might lead to court cases.

Especially the technical communication manager will help you with strongly positioning your product, based on features and benefits and demonstrate how your product is doing versus competitive products of the same league (if any). This comes down to establishing durable competitive differentiation, the base for you to get out of a rat race with competition.

R&D support of brand initiatives

This template describes the engagement from TCM/PR managers for the delivery of an external relations / PR toolkit for a typical brand initiative.

- *Technical communication and PR expertise is an integral part of brand initiatives.*
- *TCM and PR Managers work in synergy to provide the best and most relevant content.*
- *Brand confirms its communication message (claim) for print and ATL communication.*
- *Standard documents (musts – minimum) for initiatives to be delivered by TCM:*
    - *press releases: consumer, retail, vet (if required);*
    - *technical rationale – how does product/active ingredient work, Q&A's;*
    - *credentialing support from Key Opinion Leaders (KOL's) and external experts (endorsement/perceived endorsement);*
    - *sales presentation in the relevant format for retailer and consumer.*
- *Contacts with external experts/KOL's are managed by TCM. PR agency works directly with experts/KOL's after TCM has set the stage for cooperation.*
- *Required mind-set is of maximizing the potential of brand projects. In case of watch-outs alternative solutions are presented.*
- *Our thinking goes beyond preparation of communication materials. We look for ways to extend their use beyond the press. Including in-store, influencers or business building ideas.*
- *Both brand and R&D respect each other's deadlines and work priorities.*
- *TCM is fully on boarded and up to date on relevant CMK studies*

KVK: 54632730
BTW NL8538.04.849.B01

info@globalpets.community

Global Pets Community BV
PO Box 1719
3800 BS, AMERSFOORT, NL

Page 16 of 23

Does your marketing plan foresee PR events and/or publications of articles in consumer magazines (advertorials, editorials, or informative articles)? Then it will again be good to know that both functions will help you with either writing or revising the text. The objective is for all statements, claims and technical data in your article to be CLADD.

A very useful equity dissemination tool is having R&D employees giving lectures at veterinary or breeder congresses, or acting as speakers at seminars for all kinds of target groups (veterinarians, breeders, retail organizations, etc.). If these lectures or the research reports were published in scientific and/or specialist magazines, this would even create more value. Provided there is a communication plot available to make the end-customer aware of all the work that R&D of your company is doing to serve them and their pets better than competition does.

Last but not least there is the training aspect. It is so important that Marketing & Sales have a deep understanding about dog and cat nutrition and a detailed understanding of all the ins and outs of the products you sell. The same goes for any other function/person who contributes to selling the product. Hence this also includes external parties like independent distributors, retailers, veterinarians, and breeders or other stakeholders. Whilst you, as a product manager, will certainly have to cover some parts of this training (e.g. launch plan, commercial proposition, selling tools, etc.), it is usually an R&D person who will guide them through the scientific, technical, nutritional information. This of course should be adapted to the nature and the needs of the different target audiences.

A real life example

This story is written in the past tense since the company referred to has been acquired since.

The company took its mission of enhancing the wellbeing of pets very seriously and worked hard to achieve its vision of becoming the world leader in dog and cat nutrition. R&D was very involved with basic research and very important nutritional concepts were introduced, while other, faulty opinions on how dogs and cats were nutritionally managed in case of diseases. For instance, this company introduced the concept of balanced omega 6/omega 3 fatty acid ratio, the advantages of using moderately fermentable fibre sources, and spread to the world that extreme low protein levels in the food does not necessarily help dogs with renal failure, just to name a few.

Every other year this company shared with the entire world what their findings and progresses on research were. By organizing nutritional symposia, key opinion leaders and university professors from all over the world were invited for a two-day's information sharing event. The whole community could benefit from their work. Also competitors. All their findings were published in three volumes of Recent Advances in Canine and Feline Nutrition. The symposia were organized and lead by R&D, as their target group was about 200 top scientists.

KVK: 54632730
BTW NL8538.04.849.B01

info@globalpets.community

Global Pets Community BV
PO Box 1719
3800 BS, AMERSFOORT, NL

Page 17 of 23

4. Basic skills required

Let's assume that you, as a product manager, work with R&D on the most complex issue of product development. If you want to be productive you will have to acquire certain skills and knowledge upfront. The skills are associated to your specific work as a product manager (like writing a Basis of Interest), but 'knowledge' referred to in this lesson will be background knowledge not specific to the job of PM, but which will greatly enhance the smooth cooperation with the R&D function.

4.1 'Background' knowledge
Let's start with the 'background' knowledge you will need to acquire.

Basic nutrition
- You will need to have a decent understanding of the physiologic needs of dogs and cats as carnivores and what that means in terms of specific nutritional needs. Ideally, you will understand how these physiologic needs differ according to life-stage, life style, breed size (for dogs mainly) special conditions (like pregnancy), as well as the major health related issues of pets.
- Understanding the characteristics of the six nutrient groups, their value for optimal nutrition, the sources of nutrients and how these sources vary in terms of providing optimal nutrition.
- Providing energy is the overriding function of nutrition. Hence you will have to understand where this energy is coming from, how it dictates food intake and how, in fact, comparisons between diets can only be made on energy-base.

A very good, practical and 'accessible' reference manual would be the third edition of 'Canine and Feline Nutrition – a Resource for Companion Animal Professionals' by Case, Daristotle, Hayek and Raasch (Mosby, Elsevier). Another reference can be found in www.fediaf.org/self-regulation/nutrition

Product knowledge
In order for you to appreciate where your products stand versus competition and how they serve the needs of customers and their pets, you will need to have a solid knowledge of all the ins and outs of products.

Main elements to consider here are:
- Composition (the ingredients), the additives (elements of the food that are not 'ingredients'), analytical constituents (the chemical analysis of the food), feeding instructions (and the resulting daily feeding costs).
- Strengths and weaknesses of your products versus targeted competitive products. 'Targeted' means that you need to focus to those competitors who might represent a threat for your products or competitive products that you strategically focus on (to acquire more shelf space and/or market share).
- The product claims and their substantiation. You need to understand those for your products but also for competitive ones.

KVK: 54632730
BTW NL8538.04.849.B01

info@globalpets.community

Global Pets Community BV
PO Box 1719
3800 BS, AMERSFOORT, NL

Page 19 of 23

- Competitive differentiation: how unique are your products. What do they offer that competitive products do not or cannot offer? Or is it the other way around: do your products lack competitiveness? If your products have unique features or characteristics, how sustainable are they? Or can they easily be copied? Would it be easy for competitors to say 'the same but cheaper'?

All these elements will be required for you to make a critical analysis of where you stand in the market place. Without this you will not be able to develop a coherent and meaningful plan for product development or product improvement. Mastering this information, along with other knowledge intrinsic to your function as Product Manager, (like customer market knowledge including market trends), is a *conditio sine qua non* for achieving true innovation. Innovation creates an *intervention* in the market and is not to be confused with creating 'me too' products.

4.2 The regulatory framework

Bringing pet food to the market is subject to strict and extensive regulations. While it is not the aim to make you a specialist on the matter, it is nevertheless essential that you understand the basic principles and that you know how to retrieve information on the subject in a handy fashion. R&D will help you out on the details, will even take over this element of product development for you, but you need to be able to understand what's going on.

Understanding two parts of the legislation will help be of great help to you:

KVK: 54632730
BTW NL8538.04.849.B01

info@globalpets.community

Global Pets Community BV
PO Box 1719
3800 BS, AMERSFOORT, NL

Page 20 of 23

1. **Rules and regulations (and constraints) on how to bring a pet food to the market**
   Basically, this comes down to 'what needs to be on the label' and 'what cannot be on the label'. What is, or should be on the label is also related to how you communicate via other means in the market place (website, brochures, in-store communication, etc.)

2. **Claim development**
   What can you say (and what can you not say) about the features and benefits of your product. What proof do you need to have before you can make a certain claim (claim substantiation).

Of course you can check the EU legislation documents but perhaps it is, for your purpose, better to refer to the FEDIAF documents, that provide an excellent, handy, very accessible and practical guidance. They can be found on www.fediaf.org/self-regulation/labelling

4.3 10 steps to creating an optimal workflow

The below flowchart is a suggestion of steps to follow for product development. We will not dig deep into the specifics of the intrinsic PM job.

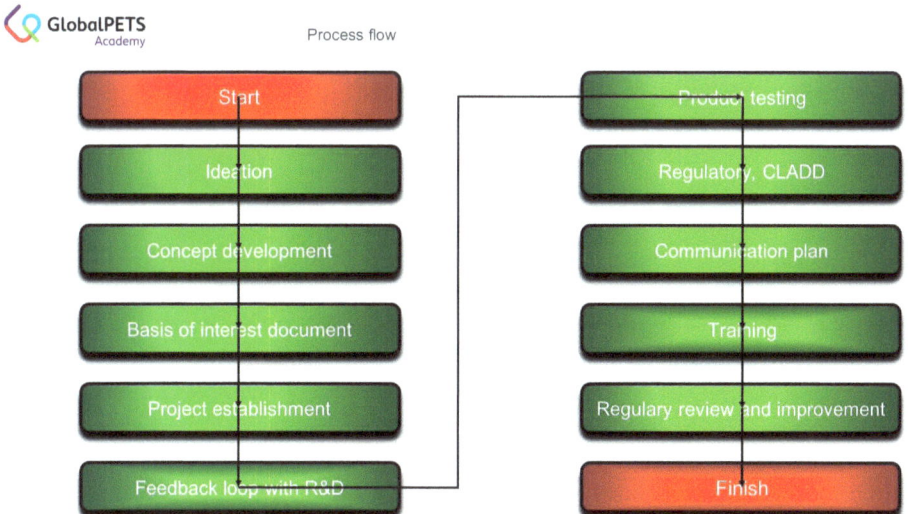

2015: © Global PETS Community BV

Step 1: Ideation

All product development starts with 'ideation': coming up with relevant ideas. Decent ideation should be based on reality. Several elements come into play in order to achieve realism. Thorough consumer marketing knowledge (CMK) should be at the base of idea creation. Competitive intelligence is a critical part of CMK. Beware that we speak here of competitive intelligence, which goes further than data gathering; rather it is the interpretation of the data.

KVK: 54632730
BTW NL8538.04.849.B01

info@globalpets.community

Global Pets Community BV
PO Box 1719
3800 BS, AMERSFOORT, NL

Page 21 of 23

Realizing true innovation is not an easy thing to do and it is a word that is often abused in the pet product world (see above). A good article on innovation and a methodology to achieve it, is 'Improving the fuzzy end of product development' by Donald Coates. It addresses continuing innovation incorporating TRIZ (theory of analysis and inventive problem solving). It is an outstanding prerequisite for the forecasting part of the next step.

Step 2: Concept development
At this stage you can already brainstorm with R&D about the initial thought. You can discuss with them whether it is 'on strategy' for product positioning, whether there is a reasonable chance that you will be able to substantiate the product features and benefits, whether there are major regulatory hurdles (and how to tackle them) or whether there are nutritional or regulatory no-go's.

Step 3: Basis of Interest document
Both previous steps will allow you to fine-tune your thoughts for the write-up of the 'Basis of Interest' document. This provides the solid base to brief management and other functions within the company, to convince them about the relevance and value of your proposal, to achieve a GO, to mobilize other functions and to acquire the budget to proceed. Thanks to analysis done in the first two steps you will be able to bring forward solid 'reasons to believe' that your proposal will lead to positive outcomes. It will allow you to make a forecasting based on the reasonable and substantiated assumptions you make and that it will result in business profit (€'s or $'s!) and/or company and brand equity (image creation with various target groups).

Step 4: Project establishment
Once you obtain the green light from management you can work on the project establishment. This document is intended to make the work to be done very concrete. You establish a cross-functional team, the objective(s), the role and tasks of the various team members and the deadlines. It is wise to assign roles and responsibilities very clearly via the RASCI model. This makes clear who is Responsible, Accountable, who is to provide Support, who has to be Consulted on regular basis and who needs to stay Informed. It avoids duplication of efforts or that things fall through the cracks. Be sure that all parties commit to the deadlines in view of other assignments they might have.

Step 5: Feedback loops
Build in regular feedback loops with R&D and keep a finger on the progress. Identify and solve upcoming hurdles and adapt the plan if necessary.

Step 6: Product testing
It is obvious that no product will be launched unless it has been demonstrated that the factory is able to deliver a quality product and that the dogs and cats respond well to the new diet. Make sure that you can evidence that the product delivers on the promises made through the product claims.

Step 7: Regulatory/CLADD
Work with R&D on the claim substantiation files and make sure that Technical Communication and Regulatory Affairs sign off on the CLADD of all claimed features and benefits.

KVK: 54632730
BTW NL8538.04.849.B01

info@globalpets.community

Global Pets Community BV
PO Box 1719
3800 BS, AMERSFOORT, NL

Page 22 of 23

Step 8: Communication plan

Establish your communication plan and make sure you book the time for active involvement of R&D people in the field if necessary.

Step 9: Training

Make sure that all internal and external training related to the new product is in place and that all parties have a profound understanding of the new product before launch.

Step 10: Regularly review and improvement

Regularly review, also with R&D, what the results are in the market place and identify possible issues to be corrected or improved.

KVK: 54632730
BTW NL8538.04.849.B01

info@globalpets.community

Global Pets Community BV
PO Box 1719
3800 BS, AMERSFOORT, NL

Page 23 of 23

www.ingramcontent.com/pod-product-compliance
Lightning Source LLC
Chambersburg PA
CBHW050913180526
45159CB00007B/2899